Alex Nogueira Pires
Renata Gizani Moura Leite

Natural Products

Alex Nogueira Pires
Renata Gizani Moura Leite

Natural Products

In the treatment of Leishmaniasis

ScienciaScripts

Imprint

Cover image: www.ingimage.com

This book is a translation from the original published under ISBN 978-613-9-67110-6.

Publisher:
Sciencia Scripts
is a trademark of
Dodo Books Indian Ocean Ltd. and OmniScriptum S.R.L publishing group

120 High Road, East Finchley, London, N2 9ED, United Kingdom
Str. Armeneasca 28/1, office 1, Chisinau MD-2012, Republic of Moldova, Europe
Printed at: see last page
ISBN: 978-620-8-14827-0

SUMMARY

SUMMARY

For centuries, man has used plants to get rid of diseases. The natural products described are secondary metabolites of various plant species, such as alkaloids, flavonoids, lignoids, among others. This study therefore describes the identification and use of natural products that are currently being studied for the treatment of leishmaniasis. The research was based on a literature review (bibliographic research) of scientific articles, in which substances extracted from the world's flora and especially from the Brazilian flora, have been showing significantly positive results for alternative treatments acting against *Leishmania.* This shows that studies are constantly evolving to find new mechanisms and alternatives for the use of natural products with less cytotoxic effects than those currently used.

Keywords: *Leishmania]* Natural products; Anti-leishmania.

1. INTRODUCTION

leishmaniasis is a disease caused by protozoa of the genus *Leishmania,* which cause different clinical pathogeneses, including cutaneous, mucocutaneous hyperergic and diffuse anergic leishmaniasis (LEON et al, 1992; 2002). The incidence is higher in regions with lower financial and social indices (DAVIES et al, 2003).

The species of *Leishmania* that circulate in certain areas where transmission is high, particularly in regions where the different clinical forms occur simultaneously, are important for understanding the epidemiology of leishmaniasis and planning control strategies (BANULS, A.; HIDE, M.; PRUGNOLLE ,F., 2007).

According to the World Health Organization (WHO, 2002), the global burden has remained stable for some years, but on the other hand it is responsible for considerable morbidity and mortality worldwide, affecting thousands of people every year in eighty-eight countries, with around three hundred and fifty million people at risk.

Natural products or medicinal plants have been studied throughout human history and traditionally used to treat diseases of different etiologies. Particularly in recent decades, phytotherapy, as studies of these products are called in relation to different discoveries in science, has received considerable attention in the search for alternatives in disease control, not only in parasitology, but also in different areas of health (MUELAS- SERRANO et al, 2000).

In Brazil, the older population living in rural areas relies on traditional and alternative medicines for the treatment of many infectious diseases (LORENZI & MATOS, 2002). Brazil's forests are home to the greatest biodiversity on the planet, thus providing a wide variety of alternative treatments for the oldest diseases (TEMPONE et al, 2005). These plants used in folk medicine have offered various natural products for the treatment of diseases caused by protozoa, so that the discovery, for example, of artemisinin, used as a pharmaceutical product for the treatment of malaria, has brought with it an interest in the discovery of new compounds for alternative treatments (CHAN-BACAB & PENA-RODRIGUES, 2001).

The biodiversity found in Brazilian forests is a potential source of many new bioactive molecules to be studied and exploited (ELISABETSKY, E. & SHANLEY, P., 1994).

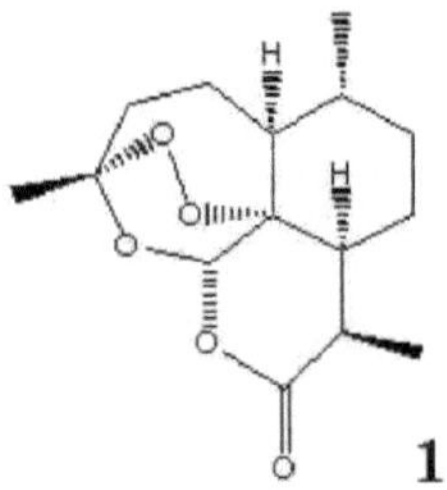

1

Figure 1: Chemical structure of artemisinin

1.1. Justification

The knowledge generated by published articles about these substances extracted from plants for medicinal use with anti-parasitic action has prompted a question. Could these molecules also act as a type of leishmanicide or alternative treatment against *Leishmania?* To this end, this literature review evaluates, observes and reports on the various molecules or substances with anti-parasitic activity and leishmanicidal action.

2. Objectives

S Report on the biodiversity of molecules extracted from plants for anti-parasitic purposes;

S Review and describe some of the molecules used as leishmanicides or with anti-leishmanial action.

3. METHODOLOGY

The method, according to Garcia (1998), represents a rational and orderly procedure (way of thinking), made up of basic instruments, which involves using reflection and experimentation, to proceed along the path (etymological meaning of method) and achieve the objectives pre-established in the research planning (project).

According to Lakatos and Marconi (1995), methods can be subdivided into approach methods and procedural methods.

According to Galliano (1986), all the meanings of the word "method" recorded in dictionaries are linked to the Greek origin "methodos" - which means "way to reach an end".

3.1. Type of research

A bibliographical study was carried out, which consists of determining an object of study, capable of serving as a body of literature for learning.

According to Gil (1991), bibliographic research can be based on published material, consisting mainly of books, journal articles and, nowadays, material available on the Internet.

3.2. Research site

The work was carried out in public and private libraries, as well as searches in databases on the World Wide Web, in order to acquire more material that was feasible for the development of this work, from August 2010 to November 2010.

3.3. Research procedures

According to Rauen (1999), it is the part that presents the results obtained in the research and analyzes them in the light of the objectives and/or hypotheses. Thus, the presentation of the data is the evidence of the conclusions and the interpretation consists of balancing the data with the theory.

For Trivinos (1987), the process of content analysis can be carried out as follows: pre-analysis (organization of the material), analytical description of the data (coding, classification, categorization), referential interpretation (treatment and reflection).

The aim of analysis is to summarize the observations so that they provide answers to the research questions. The aim of interpretation is to find the broader meaning of these answers, by linking them to other knowledge already obtained (RAUEN, 1999).

4. literature review

4.1. *Leishmania*

4.1.1. History

According to Leon (2002), since colonial times numerous scholars have provided data on a strange disease in the Andean region. In 1571, Pedro Pizarro reported that the people in the hot valleys of Peru were being decimated by a nose disease, which we now know to be leishmaniasis.

4.1.2. Gender

For Ross (1903), the genus *Leishmania* grouped together species of unicellular, heteroxenous, parasitic protozoa belonging to the order *Kinetoplastida* and *Trypanosomatidae.* They are found in promastigote flagellated forms in the digestive tract of invertebrate hosts and amastigotes, free semi-flagellates, obligate intracellular parasites of the mononuclear phagocytic system of vertebrate hosts.

Parasites of the genus *Leishmania* are flagellates of the Family *Trypanosomatidae,* Order *Kinetoplastida,* Phylum *Protozoa.* The order *Kinetoplastida* is characterized by having a single mitochondrion (called a kinetoplast) rich in mitochondrial DNA, kDNA. This organelle is located anterior to the flagellum's kinetosome, perpendicular to the organism's major axis, and has been extensively studied in the genus *Leishmania.* The morphological characteristics of the genera that make up the *Trypanosomatidae* family are very homogeneous, with a nucleus, kinetoplast, blepharoplast, flagellum and mitochondria in promastigote and amastigote forms (in relation to the *Leishmania spp genus)* (Figure 2), and all the species in this family are parasites. Irypanosomatids have a very rigid network of subpellicular microtubules, giving these protozoa well-defined cell shapes during their evolutionary cycle (CHANG et al, 1985; ROSS, 1903; WOODCOCK, 1906; HONIGBERG, 1963; VICKERMAN, 1976; SIMPSON, 1987; WILRTH etal, 1990; CHAKRABARTYet al, 1962).

4.1.3. Classification

There is enormous difficulty in classifying the species of the *Leishmania* genus, due to the great morphological similarity between them (CHANCE et al, 1979). Initially, the *Leishmania* genus had its species identified according to the clinical aspect of the diseases, (PESSOA

& PESTANA, 1941; PESSOA, 1961) although epidemiological data involving a wide variety of vectors and reservoirs suggested that this genus would certainly be composed of extremely heterogeneous parasite populations.

Based on extrinsic criteria, which tend to generate an artificial classification (LUMSDEN, 1977), such as the behavior of the parasite in culture and in hamsters (laboratory animals), Lainson and Shaw (1972) grouped the *Leishmania* species into three complexes: *Leishmania braziliensis* complex, *Leishmania mexicana* and *Leishmania donovani.* The current classification of the species in this genus is based on clinical and epidemiological characteristics, supported by biological, biochemical and molecular aspects (WHO, 1990).

Morphological differences between human parasite species were demonstrated by Shaw & Lainson (1976), using printing, fixation, staining and microphotographic techniques. More recently, several researchers have used various biochemical methods to characterize parasites, such as the fluctuation density of nuclear and kinetoplast DNA (CHANCE et al, 1974), DNA and RNA hybridization, electrophoretic analysis of kinetoplast DNA products cleaved by restriction enzymes (schizodema), electrophoretic mobility of isoenzymes (EBERT et al, 1973; GARDENER et al, 1974), radiorespirometry (DECKER-JACKSON et al, 1977), monoclonal antibodies (PRATT & DAVID, 1981) and analysis of promastigote differentiation antigens or their excretion factors (SCHNUR and ZUCKERMAN, 1977; SCHNURetal, 1972).

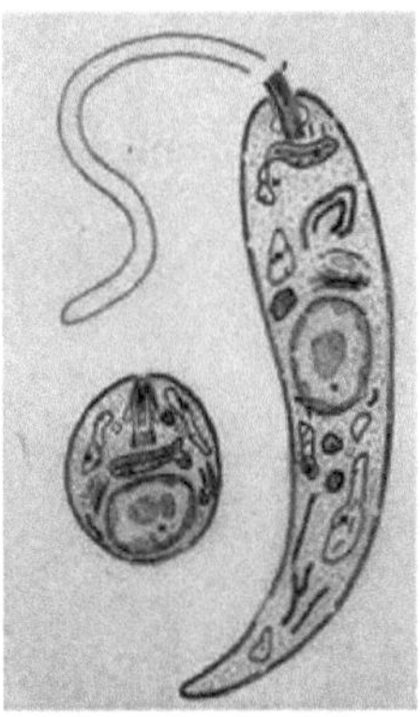

Figure 2: Leishmania morphology - schematic figure showing the nucleus, kinetoplast, nephoplast, flagellum and mitochondria of promastigotes and amastigotes. Source: FIOCRUZ, 1997.

It is currently accepted as reasonable to classify the *Leishmanias* that infect humans into phenotypic complexes, grouped into two subgenera, each comprising several species

(LAINSON & SHAW, 1987):

a) Subgenus *Viannia* (which includes the complex *braziliensis, guyanensis, lainsoni, naifi),* the *Leishmanias* show poor development in conventional culture media, slow development or visceralization in experimentally infected hamsters and develop in the posterior intestine of the phlebotomine, adhered to the wall, in the region of the pylorus (MARZOCHI, 1992);

b) Subgenus *Leishmania* (which includes the complex, *mexicana, infantum, amazonensis, enrietii, hertigi), Leishmanias* grow easily in culture, cause large nodular lesions in hamsters, with metastases to the extremities and develop in the middle and anterior intestine of the phlebotomine (MARZOCHI, 1992).

In the New World, eight species of *Leishmania* are recognized as being responsible for the disease in humans (GRIMALDI et al, 1989), belonging to the *Viannia* (V) and *Leishmania* (L) subgenera:

- *"Leishmania (V) braziliensis":* causes cutaneous and mucosal lesions with a wide geographical distribution from Central America to northern Argentina (VIANNA, 1912);

- *"Leishmania (V) guyanensis":* predominantly causes cutaneous lesions and occurs in part of South America, restricted to the Amazon Basin (FLOCH, 1954);

- *"Leishmania (V) panamensis":* predominantly causes cutaneous lesions, and occurs in Central America and the Pacific Coast of South America (LAINSON & SHAW, 1972);

- *"Leishmania (V) lainsoni":* causes skin lesions and occurs in the north of the state of Para, in the Amazon region of Brazil (SILVEIRA et al, 1987);

- *"Leishmania (L) mexicana":* causes cutaneous and eventually diffuse (anergic) lesions and occurs in Mexico and Central America (BIAGI, 1953);

- *"Leishmania (L) amazonensis":* causes cutaneous and eventually diffuse (anergic) lesions and occurs from Central America to the north, northeast and southeast of South America (LAINSON & SHAW, 1972);

- *"Leishmania (L) venezuelensis":* causes skin lesions and occurs in Venezuela (BONFANTE-GARRIDO et al, 1980);

❖ *"Leishmania (L) chagasi":* causes the visceral form with fever, anemia, hepatosplenomegaly, weight loss and occurs from Mexico to northern Argentina, predominantly in northeastern Brazil (CUNHA & CHAGAS, 1937).

The different species of the *Leishmania* genus produce a wide variety of clinical manifestations that depend on the interaction between the immune response of the vertebrate host and the invasiveness, tropism and pathogenicity of this parasite (WILSON & PEARSON, 1990).

4.1.4. Biological Cycle

According to Neves (2002), the amastigote forms of *Leishmania* are found parasitizing cells of the mononuclear phagocyte system (MPS) of the vertebrate host, mainly macrophages resident in the skin, where they multiply and survive in these cells which are specialized in destroying foreign agents. The promastigote and paramastigote forms are found in the digestive tract of phlebotomine sandflies, free or adhered to the intestinal epithelium, and the insect vector is *Lutzomyia* (Figure 3).

Parasites of the *Leishmania* genus are digenetic (heteroxenous) and have only two evolutionary forms in their life cycle: the promastigote form, which is flagellated and extracellular, and the amastigote form, which is intracellular and non-moving. The promastigotes have an elongated body measuring between 14 and 20 mm and free flagella. The amastigotes have an ovoid body measuring between 2.1 and 3.2 mm and an internal flagellum (PESSOA & MARTINS, 1982).

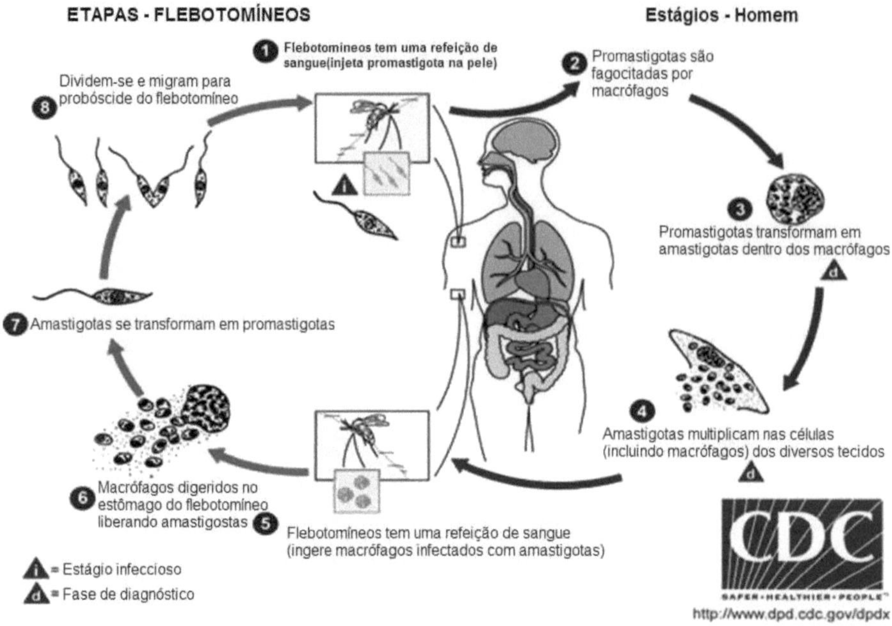

Figure 3: Leishmania Biological Cycle

4.1.5. Epidemiology

Leishmaniasis is a disease that affects approximately 12 million people worldwide. Two million new cases are diagnosed each year (WHO, 2002).

In Brazil, according to Neves (2002), more than 90% of cases are documented and detected in all the coastal states, from Para to Parana and in central states such as Minas Gerais, Goias and Mato Grosso do Sul. As the disease is typically rural, it can also be acquired in villages or suburbs of large cities - Sâo Luis (MA), Teresina (PI), Fortaleza (CE), Aracaju (SE), Rio de Janeiro (RJ), Belo Horizonte (MG), Santarém (PA), Sobral and Russas (CE), Jacobina (BA), among others - where the environmental conditions are suitable for the development of the vector.

4.1.6. Treatment

Despite its high prevalence worldwide, few advances have been made in the treatment of *leishmaniasis.* In 1912, Vianna discovered the curative action of emetic tartar, and to this day the base of this drug is used in the treatment of *leishmaniasis.* Rogers et al (1915) in India introduced trivalent antimonials for the treatment of kala-azar. Pentavalent antimonials, such as sodium stibogluconate (Pentostan) and N-methyl glucamine antimoniate (Glucantime) (Figure 2), were introduced as chemotherapy drugs in the 1940s and are currently the main drugs used to treat *leishmaniasis.*

Berman et al (1985; 1988) showed that pentavalent antimonials in the chemotherapy of *leishmaniasis* have several side effects, such as their high cardiotoxicity, very long treatment regimens and that they are not always effective.

A patient who presents lesions that are resistant to treatment does not mean that they are carrying a resistant strain of *Leishmania*, a fact that is often argued but seems difficult to prove. What seems more likely is that patients who are refractory to treatment are generally individuals whose immune status is deficient. Glucantime doesn't work until the macrophages are stimulated and the patient begins to defend himself against the parasite. In reality, immunosuppressed patients do not respond well to treatment. For this reason, the WHO has stimulated research into alternative drugs, with multicenter studies being proposed in the Mediterranean and Africa for *visceral leishmaniasis* and in South America for *tegumentary leishmaniasis,* since the association of the AIDS virus with *leishmaniasis* has increased (MACHADO et al, 1992).

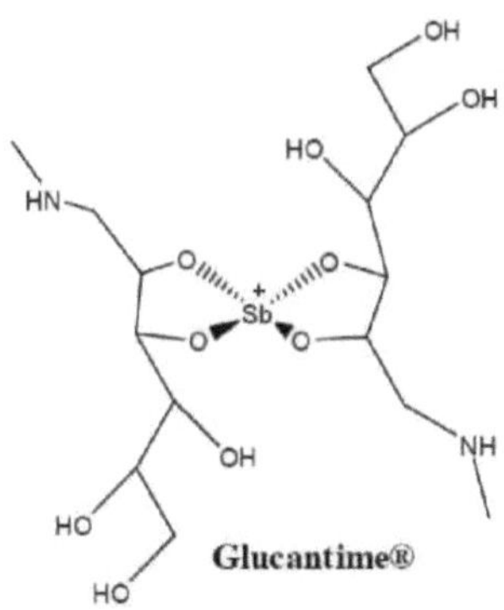

Figure 4: Proposed molecular structure of pentavalent antimonials. Meglumine antimoniate (GlucantimeS) (Frézard et al., 2008).

The drugs currently used as a second choice for this treatment are pentamidine (Figure 3) and Amphotericin B (Figure 4), which are not only very toxic, but also difficult to administer (BERMAN, 1988). The toxicity of Amphotericin B is still unclear. In addition, various compounds have been studied for their leishmanicidal activity, such as pyrimethamine and oketoconasol (NEAL & MILES, 1987), methotrexate and adriamycin (SCOTT et al, 1987).

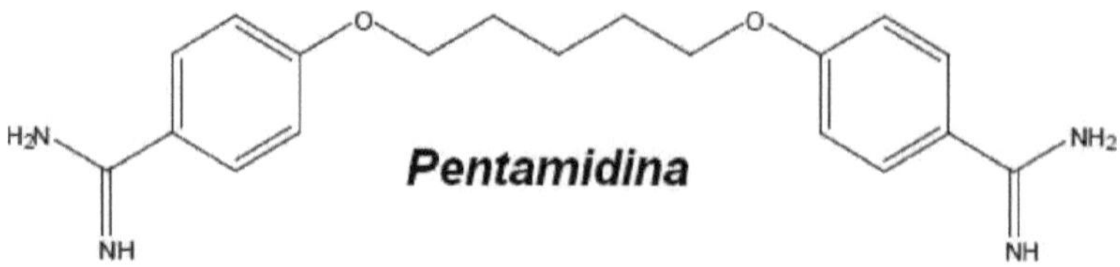

Figure 5: Molecular structure of Pentamidine (SANTOS et al, 2008).

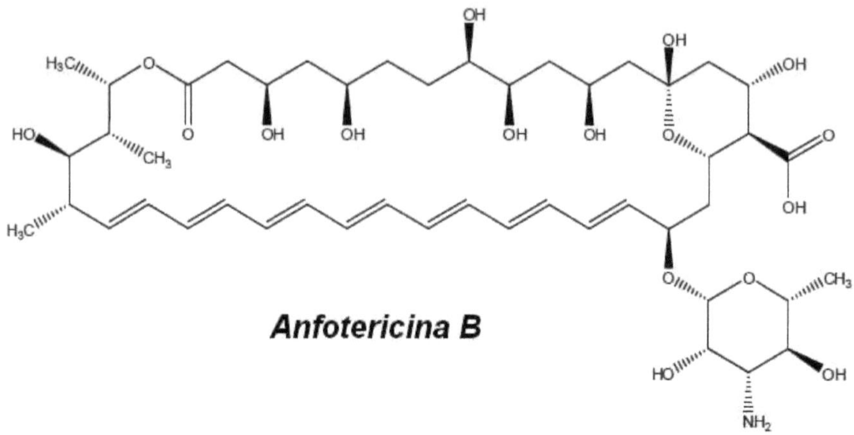

Figure 6: Molecular structure of Amphotericin B (PÉREZ-VICTORIA et al. 2006; BRAJTBURG et al., 1996)

4.2. Natural Products

The use of natural products began thousands of years ago to treat various pathologies. People in many countries today use them as an alternative and complement to synthetic medicines. Medicinal plants play an important role in world health. Despite the great advances seen in modern medicine in recent decades, they continue to be used and it is estimated that around 25% to 30% of all drugs evaluated as therapeutic agents are derived from natural products (CALIXTO, 2005; VEIGA-JUNIOR & MELLO, 2008).

In Brazil, with the rich diversity of its territory and five main biomes, the use of medicinal plants has been disseminated mainly by indigenous culture. However, this potential for discovering plants as a source of new drugs is less explored, in contrast to countries such as Germany, the United States and Canada (CALIXTO, 2000; RATES 2001;VEIGA-JUNIOR, 2008).

The regulation of the use of plants as medicines in Brazil is carried out by the National Health

Surveillance Agency (ANVISA), which, in order to guarantee the safety of the population's health, registers all medicines, a stage in which they are assessed for their safety, efficacy and quality before being offered for sale for use.

The regulation in force for the registration of herbal medicines is Collegiate Board Resolution (RDC) 48/2004, which determines the essential aspects for registration. There are also Specific Resolutions (SR): RE 88/2004, which contemplates the list of bibliographic references for evaluating the safety and efficacy of phytotherapics; RE 89/2004, which contemplates the list of simplified registration of phytotherapics; RE 90/2004, containing the guide for carrying out pre-clinical toxicity tests on phytotherapics; and RE 91/2004, which deals with the guide for carrying out alterations, inclusions, notifications and post-registration cancellations of phytotherapics (BRASIL, 2004).

In addition to the rules described above, all the legislation that regulates the registration of medicines applies to the registration of phytotherapics, among them: rules on advertising medicines, RDC 102/2000; the regulation for the certification of Good Manufacturing Practices and Control (CBPFC), RDC 210/2003 (BRASIL, 2000/2003).

Therefore, selecting a plant for pharmacological study is a very important step and can be done through traditional use, chemical components, randomized selection or a combination of more than one criterion. The most common strategy is the use of natural sources in folk medicine, which is known as ethnobotany (RATES 2001; ALBUQUERQUE & HANAZAKI, 2006).

Currently, many plants that are being studied are capable of acting on behavior, mood, thought and sensations, and understanding their mechanisms of action, safety and efficacy is a challenge for researchers (CARLINI et al, 2006).

4.3. Current Molecules Studied with Effects

Antileishmanicidal

Currently, the lack of a totally effective drug with fewer side effects for *Leishmania* has led to renewed interest in the study of traditional solutions as sources for the development of new chemotherapeutic compounds with greater activity and less toxicity. To this end, various plants have been used against parasitic diseases (ARAUJO et al, 1998).

Some molecules extracted from plants have been studied and shown to be effective as antiparasitic agents, for example: *Hydrocotyle bonariensis* (Apiaceae) "acariçoba" (LORENZI & MATOS 2002; NATARAJAN & PAILY, 1973); *Eleutherine bulbosa* (Iridaceae) "marupari" (LORENZI & MATOS, 2002); *Bauhinia forficata* (Fabaceae-Cercideae) "pata- de-vaca" (DUARTE-ALMEIDA et al, 2004 & LORENZI & MATOS, 2002) - As well as the molecules: alkaloids extracted from *Aspidosperma ramiflorum* (LORENZI, 1992); essential oils from the leaves of *Annona foetida* (Annonaceae) - among other molecules and plants that are currently being experimented on for alternative treatment.

4.3.1. Class: Alkaloids

According to Pelletier (1983), an alkaloid is a cyclic organic substance containing a nitrogen in a state of negative oxidation and whose distribution is limited among living organisms, generally with a marked biological action, such as morphine, caffeine and nicotine. Their common names, which we are more used to seeing, usually end with the suffix *ina:* caffeine, cocaine, pilocarpine, papaverine, morphine, heroin, codeine, psilocybin, among others.

They are generally white solids (with the exception of nicotine). In plants they can exist in the free state, as salts or as oxides. They also correspond to the main natural therapeutics with anesthetic, analgesic, psychostimulant, neurodepressant, etc. actions. Thus, alkaloids can be classified according to their biological activity; their chemical structure; and their blosynthetic origin (how they are produced in the plant) and are divided into three groups (PELLETIER, 1983):

- True alkaloids, which have a heterocyclic ring with a nitrogen atom and are biosynthesized by an amino acid (PELLETIER, 1983);

- Protoalkaloids, nitrogen atoms that originate from an amino acid (example: cocaine) (PELLETIER, 1983);

- Pseudo-alkaloids, which are derived from terpenes or steroids and not from amino acids (PELLETIER, 1983).

According to Araujo et al (1998), the scarcity of effective anti-leishmania drugs has led to a great deal of interest in studying traditional solutions as a source for developing new chemotherapies that are more effective and less toxic. *Aspidosperma ramiflorum* (Apocynaceae), popularly known as "guatambu", is a tree native to the forests of southeastern Brazil (LORENZI, 1992), so Reis et al (1996) comments that some alkaloids have been isolated from the bark of its stems.

Plant extracts with leishmanicidal activity are attributed to chemical compounds belonging to different groups, such as isoquinoline alkaloids, indolic alkaloids, quinones and terpenes (ARAUJO et al, 1998). According to Angenot et al (1991), the alkaloid extract of *A. ramiflorum* was chosen due to the presence of monoterpene bisindolic alkaloids (figure 5) with structures similar to alkaloids from *Strychnos usambarensis,* which has been reported to have antiprotozoal activity.

Oliveira et al (2002), describes that the main constituents of *A. ramiflorum* are: ramiflorin A and ramiflorin B (1-2 in figure 5) - The presence was monitored by High Performance Liquid Chromatology (HPLC) (Figure 6) - The efficacy of an alkaloid extracted from *A. ramiflorum* was evaluated against promastigote forms of *Leishmania amazonensis* and *Leishmania braziliensis.*

The positive preliminary studies showed that the alkaloid extract was more effective against *L. amazonensis* than *L. braziliensis,* thus suggesting further work with these compounds.

(1)

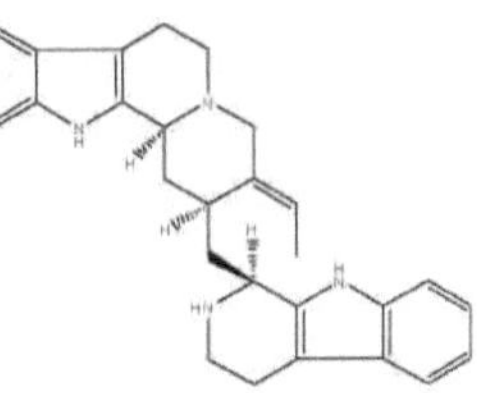

ramiflorina A

(2)

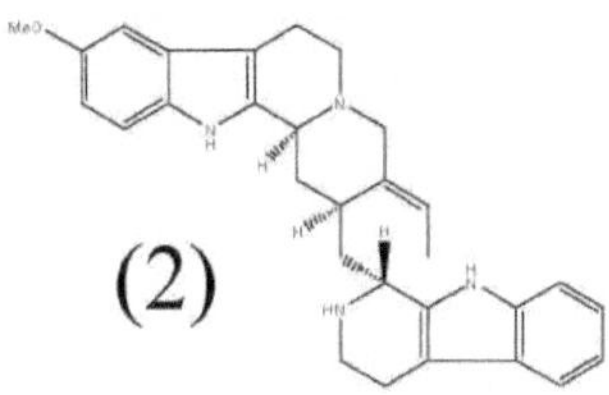

ramiflorina B

(3)

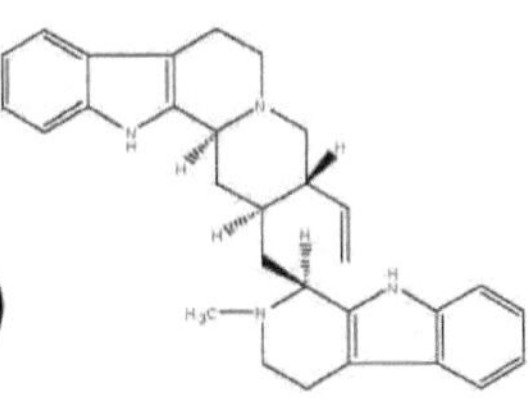

usambarina

(4)

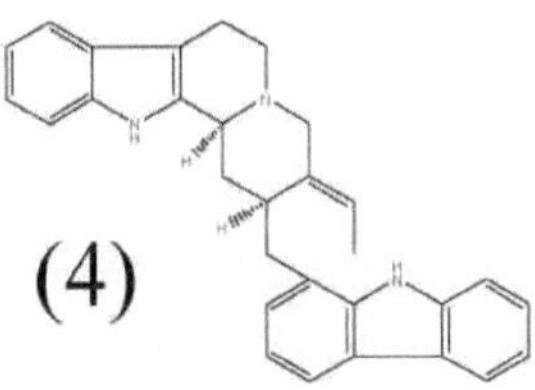

usambarensina

Figure 7: Alkaloids from *Aspidosperma ramiflorum* (1 and 2) and *Strychnos usambarensis* (3 and 4)

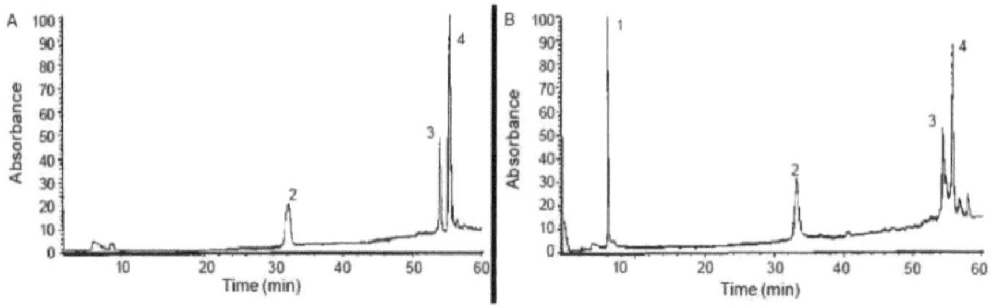

Figure 8: A: High Performance Liquid Chromatography (HPLC) - chromatogram of the mixture of standards isolated from *A. ramiflorum-* B: Chromatogram of alkaloid extracts. Peaks - 1: internal standard (tryptofol); 2: 10-methoxy-geissoschizol; 3: rhamiflorin A; 4: rhamiflorin B. Source: Oliveira et al (2002).

4.3.2. Class: Terpenoids

Terpenoids are natural hydrocarbons that can be in open or cyclic chains and are derived from isoprenes (C_5H_8), also giving rise to diterpenes and triterpenes (HARBONE & BAXTER, 1995).

The *Annonaceae* family are tropical and subtropical trees with a number of studies and medicinal properties. Pharmacological studies have shown important bioactive compounds, such as antibiotics, insecticides and antiparasitic agents - especially those acting against *Leishmania sp. Annona foetida,* popularly known as "graviola do mato" has anti-parasitic activity, for which these activities are attributed to: aporphins, β-carboline, alkaloids, lactones and essential oils (CHATROU et. al., 2004; KESSLER et. al., 1993; CORREA, 1984; KHAN et. al., 2002; RAHMAN et. al., 2005; COSTA et. al., 2006; QUEIROZ et. al., 1996; MONTENEGRO et. al., 2003; AKENDENGUE et. al., 2002; SAHPAZ et. al., 1994; NKUNYA et. al., 1993; BOYOM et. al., 2003; RIBEIRO et. al., 1999).

According to Chan-Bacab & Pena-Rodriguez (2001) and Rocha et. al. (2005), the evaluation of the leishmanicidal activity of *A. foetida* oil presented in Table 1 shows that this oil was active against all the *Leishmania* species tested, so that the best result was against *L. guyanensis* with an IC_{50} value of 4.1 µg/mL. The oil was also tested for cytotoxicity using parasitized hamster rats and Balb/c mice, which showed low cytotoxicity.

Table 1: In vitro activity of *Annona foetida* essential oil against *Leishmania* species (promastigotes)

according to Clso

Sample	L. amazonensis	L. braziliensis	L. chagasi	L. guyanensis
Essential oil	16.2+ or - 1.9	9.9 + or - 1.2	27.2 + or-6.2	4.1 + or-0.2
Pentamidine	4.8 + ou-0.1	1.0 + ou-0.1	19.0 + ou-2.9	0.3 + or-0.1

4.3.3. Class: Quinones

Quinones are one of the two isomers of the $C_6H_4O_2$ cyclohexanedione position. Quinones, particularly hydroxyquinones and their derivatives, have been studied against various parasite species (CROFT et al, 1992; FERREIRA et al, 2002; HAZRA et al, 2002), and according to Kayser et al (2000), many naphthoquinones have shown activity against *leishmania.*

According to Evani et al (2002), lapachol is a naphthoquinone isolated from various plants of the Bignoniacea family, popularly extracted from the "Ipê roxo" *(Tabeburia avellanedae)* and easily found in the North and Northeast regions of Brazil. It is known for its anti-inflammatory, analgesic, antibiotic, anti-malarial, anti-trypanosome and anti-ulcer activities.

According to Teixeira et al (2001), lapachol was effective *in vitro* against promastigote forms of *L. braziliensis.* Evaluating the efficacy of the compounds (lapachol, isolapachol, lapachol potassium salt, isolapachol potassium salt, acetylisolapachol, dihydrolapachol and petamidine isethionate) in Table 2 against *L. amazonensis* and *L. braziliensis,* the compounds showed activity against promastigote forms in the CI50/24h range of 1.6 - 7.8 µg/mL in *L. amazonensis* and 3.4 - 54.0 µg/mL in *L. braziliensis. amazonensis* and 3.4 - 54.0 µg/mL in *L. braziliensis* (BARRAL et al, 1991; LEON et al, 1992), also showing that among the lapachol analogues tested, the one with the greatest activity against *Leishmania* was acetylisolapachol (5 - Table 2).

Table 2: leishmanicidal activity (CI50), in µg/mL, of assays of

	IC_{50} µg/ml	
Compounds	*Leishmania amazonensis*	*L. braziliensis*
Lapachol (1)	5.2 ± 0.7	11.9 ± 6.9
Isolapachol (2)	4.4 ± 2.9	9.3 ± 2.7
Lapachol. K^+ salt (3)	7.7 ± 4.1	21.4 ± 2.9
Isolapachol. K^+ salt (4)	7.8 ± 0.2	15.8 ± 0.0
Acetylisolapachol (5)	1.6 ± 0.0	3.4 ± 0.5
Dihydrolapachol (6)	7.3 ± 0.3	54.0 ± 9.0
Pentamidine Isethionate	0.28 ± 0.05	11.6 ± 1.6

hydroxoquinones and their derivatives.

Source: Nadja et al, 2004

In terms of toxicity, Duke (2002) said that studies have shown that hydroxyquinones are hemolytic agents in dogs and monkeys, but in humans toxicity has been reported at doses above 1.5 g/day.

4.3.4. Class: Lignoids

According to Gottlieb (1988), *lignoids* are micromolecules whose skeleton is formed exclusively by the phenyl-propanic group and have pharmacological activities in humans, the most numerous groups of which are the *lignans.* According to Hartwell and Scherecker (1958), *lignans* fall into three classes: dibenzylbutanes, phenyltetanes and dibenzocyclooctane - so these *lignans* have antioxidant, antitumor, antiviral and analgesic activities (CHEN et al, 2002; GONZALES et al, 1990; BEDOWS & HATFIELD, 1982; BORSATO et al, 2000).

According to Morais et al (2009), the *lignoids* (1,2 and 3- Figure 7) obtained from *I. juruensis* arils showed leishmanicidal activity against the promastigote forms of *L. amazonensis* much more effectively than the other *lignoids* tested.

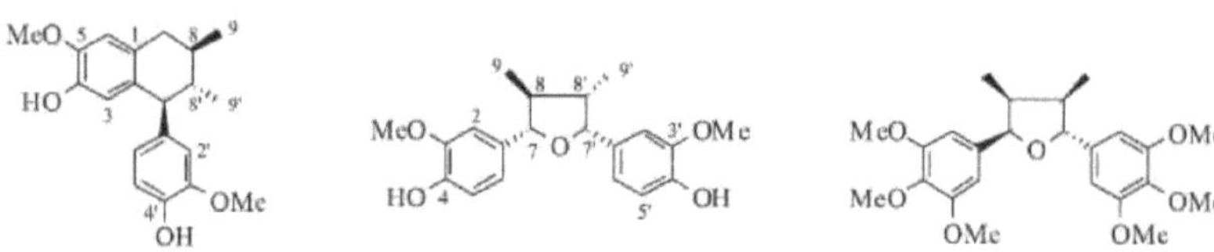

1 2 3

Figure 9: Molecular structures of lignoids; Source: Morais et al, 2009

According to Barata et al (2000), leishmanicidal activity has been reported for some species of *lignans* and some synthetic analogues, such as dihydrobenzofurans and benzofurans.

Dihydrobenzofurans and benzofuran derivatives were evaluated against amastigote forms of *Leishmania donovani* and their results showed significant efficacy (Miert et al, 2005).

4.3.5. Class: Flavonoids

According to Zuanazzi et al (2004) and Valderrama (2000), *flavonoids* are the most diverse group of polyphenolic compounds among natural products. Their structure includes two aromatic nuclei coupled by a bridge of three carbon atoms (C6 - C3 - C6), where the oxidation state of the C3 unit differentiates the types of flavonoids (Figure 8).

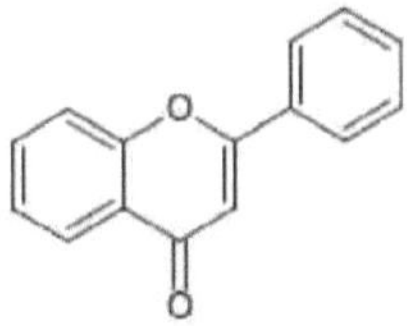

Flavonas

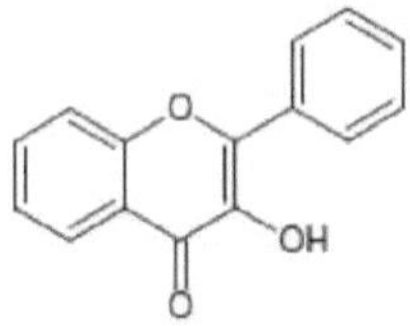

Flavonóis

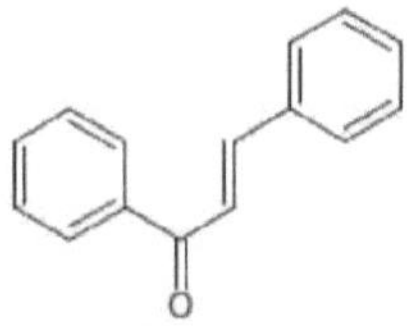

Chalconas

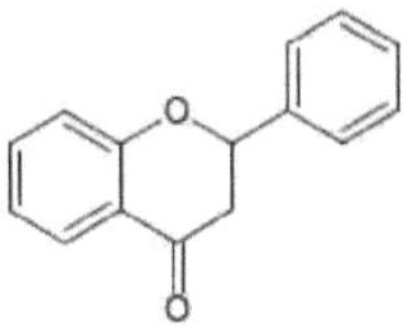

Flavanonas

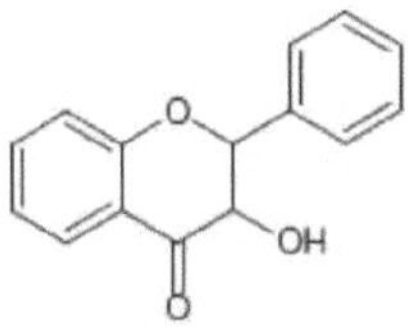

Flavanonóis

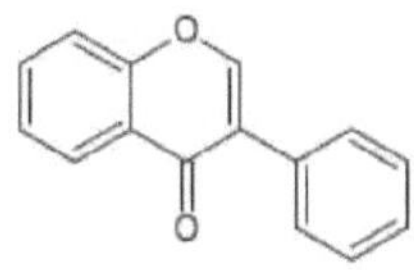

Isoflavonas

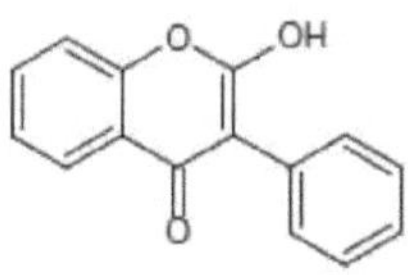

Isoflavonóis

Figure 10: Chemical Structure of *Flavanoids*. Source: Zuanazzi et al (2004) and Valderrama (2000).

According to Lin et al (1997), *flavonoids* have the following activities: antioxidant, anti-inflammatory, vasodilator, anti-allergic, anti-tumor, anti-ulcerogenic, anti-microbial and anti-viral.

Flavonoids have been reported in studies to have a broad pharmacological aspect, acting actively as: antioxidant, antimicrobial, anti-inflammatory, antitumor, antiprotozoal, antidiabetic and chemopreventive (KEMERTELIDZE et al, 2000; HIROYUKI et al, 1998; ABAD et al, 1993; COWARD et al, 1993; LUNARDI et al, 2003; SHIN et al, 1999; CASSADY, 1990).

Borges-Argaez et al (2000, 2002) reported the isolation and identification of a new series of *flavonoids* extracted from *Lonchocarpus xuul* and *Lonchocarpus yucatanensis* (Figure 9).

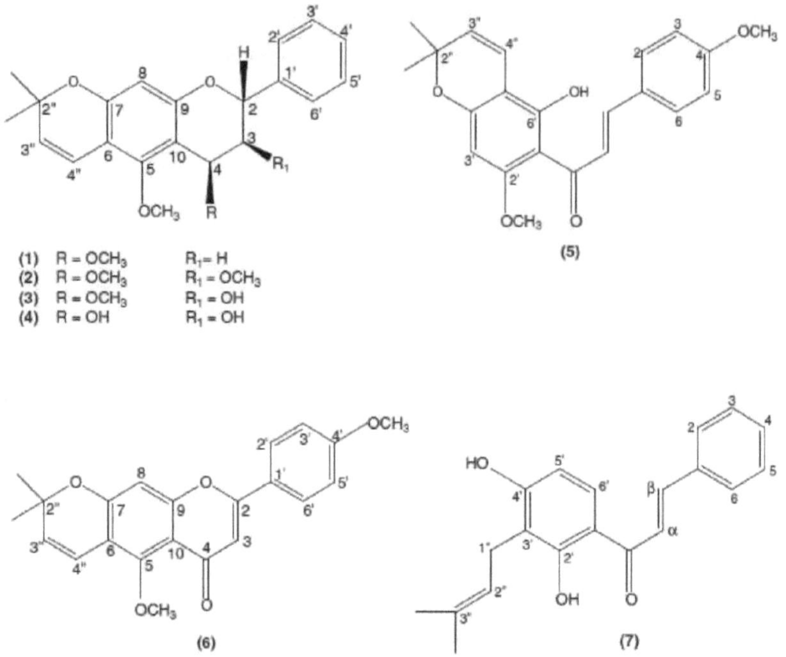

Figure 11: Structures of *flavonoids* recently reported from *Lonchocarpus spp.* Source: Borges-Argaez et al, 2007

These new *flavonoids* have been tested *in vitro* with the aim of reporting their leishmanicidal, antimalarial, trypanocidal and cytotoxic activities.

Of the metabolites evaluated, *flavonoids* 3, 4, 6 and 7 (Figure 9) showed leishmanicidal activity, where *flavonoid* 6 (Figure 9) was the most active among the others tested (BORGES-ARGÂEZ et al, 2007).

In 2006, Muzitano et al. evaluated the leishmanicidal activity of *flavonoids* extracted from *Kalanchoe pinnata.*

Kalanchoe pinnata is a medicinal herb popularly used in Brazil for the treatment of inflammatory diseases. Recent studies on the chemical composition of *K. pinnata* have shown that its main metabolites are: *bufadienolides, terpenoids and flavonoids* (ROSSI-BERGMANN et al, 1994; YAMAGISHI et al, 1989; COSTA et al, 1995).

According to Muzitano et al (2006), *K. pinnata* has great phytotherapeutic potential, as shown by its activities: anti-inflammatory and anti-ulcerogenic (PAL & CHAUDHURI, 1991), hepatoprotective (YADAV & DIXIT, 2003), anti-leishmania (DA SILVA et al, 1995).

According to Schinor et al (2006), the activities of *Chresta exsucca* extracts had the best results for leishmanicidal activity, which was due to *flavonoids*, among other compounds identified. These extracts significantly interfered with the viability of *L. amazonensis* amastigotes, where hexanic, dichloronetanic and ethanolic extracts reduced the parasite's viability to 1.8, 4.7 and 0.2%.

4.3.6. Class: Triterpenoids

According to Morais (2008), the *terpene* class derives from the condensation of isoprene units and is classified according to the number of these units present in the carbon skeleton, where they are widely distributed in nature, especially in the plant kingdom.

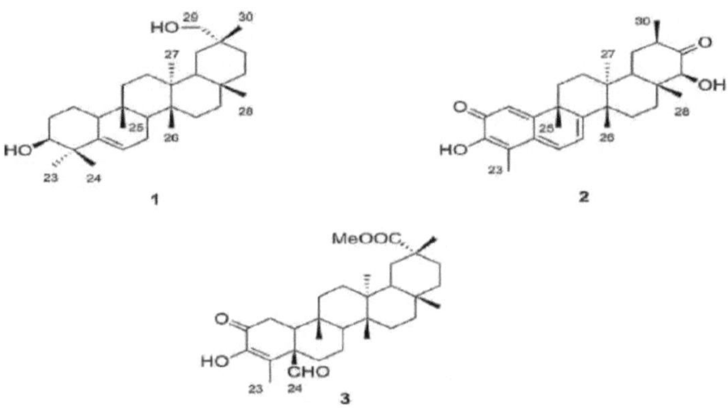

Figure 12: Chemical structure of -triterpenics Source: Maregesi et al, 2010.

Maregesi et al (2010), selected *Elaeodendron schlechteranum* for an in-depth investigation based on some preliminary results previously studied. *E. schlechteranum* are shrubs belonging to the Celastraceae family, widely used by herbalists and traditional healers in Tanzania for the treatment of anemia, dysmenorrhea, infertility, cardiovascular problems, hypertension, inflammation and antibacterial (TONA et al, 1999, 2001, 2004; MAREGESI et al, 2007,2008; WATT & BREYER-BRANDWIJK, 1962).

Maregesi et al (2010) extracted an active triterpene (compound 2 - Figure 10) and two inactive ones (compounds 1 and 3 - Figure 10), where antiparasitic activities were observed against *Trypanosoma cruzi, Trypanosoma brucei, Plasmodium falciparum* and *Leishmania infantum* with CI_{50} values of: < 0.25, < 0.25, 0.36 and 0.51ug/mL, respectively.

The triterpenes showed cytotoxic, antimicrobial, antiparasitic, insecticidal and anti-inflammatory activities (THIEM et al, 2005; BRINKER et al, 2007; MURAYAMA et al, 2007; LINDSEY et al, 2006).

4.3.7. Class: Phenolic compounds

According to Shahidi & Naczk (1995), phenolic compounds are defined as substances that have an aromatic ring with one or more hydroxyl substituents and are divided into classes.

One of these classes includes the subclasses of benzoic acids (gallic acid, protocatechuic acid, p-hydroxybenzoic acid), hydroxycinnamic acids (coumaric acid, caffeic acid, ferulic acid), hydrolyzable tannins, lignin, among others. In another class are flavonoids, proto-anthocyanins, anthocyanidins and isoflavones (FARAH & DONANGELO, 2006).

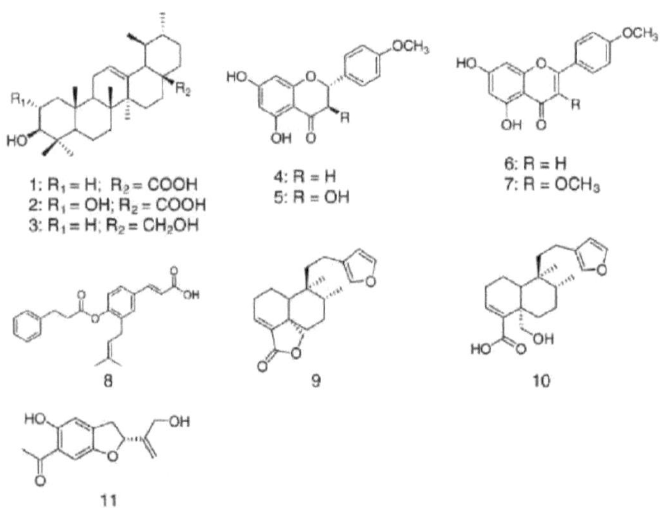

Figure 13: Chemical structures: Ursolic acid (1); 2α-hydroxyursolic acid (2); uvaol (3), isosakuranetin (4); aromadendrin-4'-methylether (5); acacetin (6); ermanine (7); baccharin (8); hautriwaic acid lactone (9); clerodane diterpenes (10); viscidone (11). Source: Da Silva Filho et al. 2009.

Baccharis dracunculifolia, popularly known as "rosemary of the field" and "broom", is used in traditional medicine as an anti-inflammatory and for gastrointestinal diseases (DA SILVA FILHO et al, 2004; SOUZA et al, 2007; LEMOS et al, 2007). Propolis is a natural substance that bees collect from plant buds and exudates (JORGE et al, 2008; SIMOES et al, 2004).

Reports state that *B. dracunculifolia* and propolis have anticariogenic, antimicrobial, antiulcerogenic, immunomodulatory, trypanocidal and leishmanicidal activities (LEITÂO et al, 2004; DA SILVA FILHO et al, 2008; LEMOS et al, 2007; MISSIMA et al, 2007; DA SILVA FILHO et al, 2004; PONTIN et al, 2008).

According to Da Silva Filho et al (2009), the phenolic compound metabolites evaluated showed leishmanicidal activity *in vitro* with CI_{50} values of 45 µg/mL, so that compounds 1 and 9 exhibited the highest leishmanicidal activities with CI_{50} values of 3.7 µg/mL and 7.0 µg/mL, respectively. As shown in the table below:

Table 3: *In vitro* activities: leishmanicidal, antimalarial and cytotoxic of *B. dracunculifolia* constituents

Extracts/ compounds	*L. donovani* (µg/mL)		*P. falciparum* (µg/mL)				Cytotoxicity (Vero cells)
	IC_{50}	IC_{90}	D6 clone[a]		W2 clone[b]		
			IC_{50}	SI[c]	IC_{50}	SI[c]	IC_{50} (µg/mL)
BdE	45	93	25	>1.9	13	>3.7	NC
GPE	49	92	20	>2.4	13	>3.7	NC
1	3.7	6.8	1.0	>4.8	NT	-	NC
2	19.0	34.0	3.2	>1.5	3	>1.6	NC
3	15.0	33.0	3.3	>1.4	1.9	>2.5	NC
4	NA	NA	NA	-	NT	-	NC
5	NA	NA	NA	-	NT	-	NC
6	18.0	40.0	NA	-	NT	-	NC
7	40.0	>40	2.6	>1.8	2.2	>2.2	NC
8	NA	NA	NA	-	NT	-	NC
9	7.0	28.0	0.8	>5.6	2.2	>2.2	NC
10	NA	NA	3.0	>1.6	2.6	>1.8	NC
11			1.9	>2.5	2.3	>2.1	NC
Chloroquine	-	-	0.018	>556	0.176	>56	NC
Artemisinin	-	-	0.014	>669	0.008	>1176	NC
Pentamidine	1.9	8.5	-	-	-	-	-
Amphotericin B	0.7	1.7	-	-	-	-	-

NA = **non-active (concentration of pure compounds: P. falciparum 4.7 mcg / mL; L. donovani 40 mcg / mL);** NT = **not tested;** NC = **non-cytotoxic (up to the maximum dose tested);** a **- A chloroquine-sensitive clone;** b **- chloroquine-resistant clone;** c - **Selectivity Index = CI_{50} Vero cells/CI_{50} P. falciparum.** Source: Da SilvaFilho et al, 2009

Some species of the genus *Mitracarpus* (Rubiaceae) have pharmacological activity widely used in traditional medicine in West Africa against headaches, toothache, amenorrhea, dyspepsia, antifungal and antimicrobial (BISIGNANO et al, 2000; MOULIS et al, 1992; SANOGO et al, 1996; SCHLECHT et al, 2007).

According to Fabri et al (2009), the phenolic compounds extracted from *M. frigidus* were active against *Leishmania amazonensis* (CI_{50} of 58.5, 11.8 and 40.0 µg/mL) and *Leishmania chagasi* (CI_{50} of 6.7 and 14.8 µg/mL).

4.4. Final considerations

This bibliographical study has shown some of the main molecules extracted from various natural products that act against different types of *Leishmania,* a parasite that is prevalent all over the world and in several endemic areas in Brazil. Therefore, this work was based on reporting and describing these molecules so that they can be used both for scientific knowledge and for the general population, who have an interest in knowledge and who are curious to use more and more of these natural products in their research.

5. References

Abad, M. J.; Bermejo$_1$ P.; Villar, A.; Valverde, S. *Antiinflammatory activity of two flavonoids from Tanace microphyllum.* Journal of Natural Products. 56, 1164, 1993;

Akendengue, B.; Roblot, F.; Loiseau, P. M.; Bories, C.; Ngou-Milama, E.; Laurens, A.; Hocquemiller, R. *Klaivanolide, an antiprotozoal lactone from Uvaria klaineana.* Phytochemistry. 59, (8):885-8, 2002;

Albuquerque, U. P.; Hanazaki, N. *Ethnodirected research in the discovery of new drugs of medical and pharmaceutical interest: Fragility and perspectives.* Revista Brasileira de Farmacognosia, v. 16, p. 678-689, 2006;

Angenot L, Quentin-Leclercq J, Phillipson DJ, Warhurst DC, O'Neill MJ, Bray DH, Wright CW. *Antiamoebic and antiplasmodial activities of alkaloids isolated from Strychnos usambarensis.* Planta Medica, 57: 337-340, 1991;

Araujo C. A. C.; Alegrio L. V.; Leon L. L. *Antileshmanial activity of compounds extracted and characterized from Centrolobium sclerophyllum.* Phytochemistry, 49: 751-754, 1998;

Banuls A.; Hide M.; Prugnolle F. *Leishmania and the Leishmaniases: A Parasite Genetic Update and Advances in Taxonomy, Epidemiology and Pathogenicity in Humans.* Advances in Parasitology, 64:1-109, 2007;

Barata, L. E. S.; Santos, L. S.; Ferri, P. H.; Phillipson, J. D.; Paine, A.; Croft, S. L. *Anti-leishmanial activity of neolignans fromVirola species and synthetic analogues.* Phytochemistry, 55: 589-595, 2000;

Barral, A.; Pedral-Sampaio, D.; Grimaldi Jr, G.; Momen, H.; Mc-Mahon-Pratt, D.; Ribeiro de Jesus, A.; Almeida, R.; Badaro, R.; Barral-Neto, M.; Carvalho, E. M.; Johnson Jr, W. D. *Leishmaniasis in Brazil: evidence that Leishmania amazonensıs produces a wide spectrum of clinical disease.* American Journal of Tropical Medicine and Hygiene, 44: 536-546. 1991;

Bedows$_1$ E. & Hatfield, G. M. *An investigation of the antiviral activity of Podophyllum peltatum.* Journal of Natural Products, v. 45, p. 725 - 729, 1982;

Berman, J. D. *Chemotherapy for leishmaniasis: biochemical mechanisms, clinical efficacy, and future strategies.* Reviews of Infectious Diseases, May-Jun;10(3):560-586, 1988;

Berman, J. D.; Gallalee, J. V. & Best, J. M. *Sodium stibogluconate (Pentostam) inhibition of glucose catabolism via the glycolytic pathway, and fatty acid betaoxidation in Leishmania Mexicana amastigotes.* Biochemical Pharmacology, 15;36(2):197-201, 1988;

Berman, J. D.; Waddell, D.; Hanson, B. D. *Biochemical mechanisms of the antileishmanial activity of sodium stibogluconate.* Antimicrobial Agents and Chemotherapy, 27:916-920, 1985;

Biagi, F. F. *Algunos comentarios sobre las leishmaniasis y sus agentes etiológicos, Leishmania tropica mexicana, nueva subspecie.* Medicina (México), 33:401-6, 1953;

Bisignano, G.; Sanogo, R.; Marino, A.; Aquino, R.; D'Ângelo, V.; Germano, M. P.; De Pasquale, R.; Pizza, C. *Antimicrobial activity of Mitracarpus scaber extract and isolated constituents.* Letters in Applied Microbiology, 30, 105-108, 2000;

Bonfante-Garrido, R.; Morillo, N.; Torres, R. *Canine cutaneous leishmaniasis in Venezuela.* Boletin de la Oficina Sanitaria Panamericana, *91:*161-165, 1980;

Borges-Argaez, R.; Balnbury, L.; Flowers, A.; Giménez-Turba, A.; Ruiz, G.; Waterman, P. G.; Pena-Rodriguez, L. M. *Cytotoxic and antiprotozoal activity of flavonoids from Lonchocarpus spp.* Phytomedicine, 14, 530-533, 2007;

Borges-Argaez$_1$ R.; Pena-Rodriguez, L. M.; Waterman, P. G. *Flavonoids from the stem bark of Lonchocarpus xuul.* Phytochemistry, 54, 611-614, 2000;

Borges-Argaez, R.; Pena-Rodriguez, L. M.; Waterman, P. G. *Flavonoids from two Lonchocarpus species of the Yucatan Peninsula.* Phytochemistry, 60, 533540, 2002;

Borsato, M. L. C.; Grael, C. F. F.; Souza, G. E. P. & Lopes, N. P. *Analgesic activity of the lignans from Lychnophora ericoides.* Phytochemistry, v. 55, p. 809 -813, 2000;

Boyom, F. F.; Ngouana. V.; Zollo, P. H. A.; Menut, C.; Bessiere, J. M.; Gut, J.; Rosenthal, P. J. *Composition and anti-plasmodial activities of essential oils from some Cameroonian medicinal plants.* Phytochemistry, Volume 64, Number 7, December, pp. 1269-1275(7), 2003;

Brajtburg, J. & Bolard, J. *Carrier effects on biological activity of amphotericin B.* Clinical Microbiology Reviews, 9:512-31. 1996;

Brazil, *Resolution N 340.* National Health Council - CNS. 2004;

Brazil. *Resolution of the Collegiate Board of Directors. RDC Resolution No. 33,* 2003;

Brinker, A. M.; Ma, J.; Lipsky, P. E.; Raskin, I. *Medicinal chemistry and pharmacology of genus Tripterygium (Celasteraceae).* Phytochemistry, 68, 732-766, 2007;

Calixto, J. B. *Efficacy, safety, quality control, marketing and regulatory guidelines for herbal medicines (phytoterapeutic agents).* Brazilian Journal of Medical and Biological Research, 33: 179-189, 2000;

Calixto, J. B. *Twenty-five years of research on medicinal plants in Latin America: a personal review.* Journal of Ethnopharmacology, *100:* 131-134, 2005;

Carlinil1 Elisaldo Araujo et. al. *II Levantamento domiciliar sobre o uso de drogas psicotrópicas no Brasil: estudo envolvendo as 108 maiores cidades do país:* Sao Paulo: CEBRID - Centro de Informações Sobre Drogas Psicotrópicas: UNIFESP - Universidade Federal de Sao Paulo, 2005, 2006;

Cassady, J. M. *Natural products as a source of potential cancer chemotherapeutic and chemopreventive agents.* Journal of Natural. Products. Vol. 53, p. 23, 1990;

Chakrabarti, G.; Basu, A.; Manna, P. P.; Mahato, S. B.; Mandal, N. B.; Bandyopadhyay, S. *Indolylquinoline derivatives are cytotoxic to Leishmania donovani promastigotes and amastigotes in vitro and are effective in treating murine visceral leishmaniasis.* Journal of Antimicrobial Chemotherapy, vol. 43, issue 3, p. 359-366, 1962;

Chan-Bacab, M. J.; Pena-Rodriguez, L. M. *Plant natural products with leishamanicidal activity.* Natural Product Reports, *18:* 674-688, 2001;

Chance, M. L.; Peters, W. and Shchory L. *Biochemical taxonomy of Leishmania. I: Observations on DNA.* Annals of Tropical Medicine and Parasitology, 68, p. 307-316, 1974;

Chance, M. L.; Peters, W.; Shchory, L. *Biochemical taxonomy of Leishmania. I. Observations on DNA.* Annals of Tropical Medicine and Parasitology, 1979;

Chang, C. S. & Chang, K. P. *Heme requirement and acquisition by extracellular and intracellular stages of Leishmania mexicana amazonensis.* Molecular and Biochemical Parasitology, 16:267-276, 1985;

Chatrou, L. W.; Rainer, H.; Maas, P. J. M.; Smith, N.; Mori, S. A.; Henderson, A.; Stevenson, D. W.; Heald, S. V. *In Annonaceae (Soursop Family).* Princeton University Press: New

Jersey, p. 18-20, 2004;

Chen, D.; Zhang, S.; Kozuka, M.; Sun, Q.; Feng, J.; Wang, Q.; Mukaianaka, T.; Nobukuni, Y.; Tokuda, H.; Nishino, H.; Wang, H.; Natschke, S. & Lee, K. *Interiotherins C and D, Two New Ligans from kadsura interior and AntitumorPromoting Effects of Related Neolignans on Epstein-Barr Virus Activation.*

Journal of Natural Products, v. 65 (9), p. 1242 - 1245, 2002;

Corrêa, M. P.; *Dicionario das plantas ùteis do Brasil, Ministério da Agricultura.* Instituto Brasileiro de Desenvolvimento Florestal, vol. 1, p.154, 1984;

Costa, E. V.; Pinheiro, M. L. B.; Xavier, C. M.; Silva, J. R. A.; Amaral, A. C. F.; Souza, A. D. L.; Barison, A.; Campos, F. R.; Ferreira, A. G.; Machado, G. M. C.; Leon, L. L. P. J. *A Pyrimidine-β-carboline and Other Alkaloids from Annona foetida with Antileishmanial Activity.* Journal of Natural Products, *69* (2), pp 292-294, 2006;

Costa, S. S.; Jossang, A.; Bodo, B. In: *Allorge-Boiteau, Kalanchoe (Crassulacées) de Madagascar, Systématiqu.* Ecophysiologie et Phytochimie, vol. 1, Paris, p. 219, 1995;

Coward, L.; Barnes, N.; Setchell, K.; Barnes, S. *Genistein, Daidzein, and their b-glycoside conjugates: antitumor isoflavones in soybean foods from American and Asian diets.* Journal of Agricultural and Food Chemistry, 41, 1961-1967, 1993;

Croft, S. L.; Hogg, J.; Gutteridge, W. E.; Hudson, A. T.; Randall, A. W. *The activity of hydroxynaphthoquinones against Leishmania donovani.* Journal of Antimicrobial Chemotherapy, 30: 827-832, 1992;

Cunha, A. M.; Chagas, E. Nova espécie de protozoario do gènero Leishmania patogênico para o homem (nota prèvia). 0 Hospital (Rio de Janeiro), Rio de Janeiro, v.11, n.2, p. 5-9, 1937;

Da Silva Filho, A. A.; Bueno, P. C. P.; Gregorio, L. E.; Silva, M. L. A.; Albuquerque, S.; Bastos, J. K. J. *In vitro trypanocidal activity evaluation of crude extract and isolated compounds from Baccharis dracunculifolia DC (Asteraceae).* Journal of Pharmacy and Pharmacology, *56:1195-99, 2004;*

Da Silva Filho, A. A.; Resende, D. 0.; Fukui, M. J.; Santos, F. F.; Pauletti, P. M.; Cunha, W. R.; Silva, M. L. A.; Gregorio, L. E.; Bastos, J. K.; Nanayakkara, N. P. D. *In vitro antileishmanial, antiplasmodial and cytotoxic activities of phenolics and triterpenoids from Baccharis dracunculifolia D. C. (Asteraceae).* Fitoterapia, 80, 478-482, 2009;

Da Silva Filho, A. A.; Sousa, J. P.; Soares, S.; Furtado, N. A.; Andrade e Silva, M. L.; Cunha, W. R.; Gregorio, L. E.; Nanayakkara, N. P.; Bastos, J. K. Zeitschrift für *Naturforschung, 63, 40-46, 2008;*

Da Silva, S. A. G.; Costa, S. S.; Mendonça, S. C. F.; Silva, E. M.; Moraes, V. L. G.; Rossi-Bergmann, B. *Therapeutic effect of oral Kalanchoe pinnata leaf extract in murine leishmaniasis.* Acta Tropica, 60, 201- 205, 1995;

Davies, C. R.; Kaye, P.; Croft, S. L.; Sundar, S. *Leishmaniasis: new approaches to disease control.* British Medical Journal, 326:377-382, 2003;

Decker-Jackson, J. E.; Schrot, J. R.; Levin, G. V. *Identification of Leishmania spp by radiorespirometry.* Journal of protozoology, 24(3):463-70, 1977;

Duarte-Almeida, J. M.; Negri, G.; Salatino, A. *Volatile oils in leaves of Bauhinia (Fabaceae-Caesalpinioidae).* Biochemical Systematics and Ecology, 32:747753, 2004;

Duke, J A.; Press, C. R. C.; Raton, F. L. B. *Handbook of Medicinal Herbs.* ed. 2; 2002;

Ebert, F.; Buse, E.; Mühlpfordt, H. *In vitro light and electron microscope studies on different virulent promastigotes of Leishmania donovani in hamster peritoneal macrophages.* Zeitschrift für Parasitenkunde, 1973;

Elisabetsky1 E. & Shanley, P. *Ethnopharmacology in the Brazilian Amazon. Pharmacology and Therapeutics.* 64: 201-214, 1994;

Evani, L. A.; Joao, R. B. A.; Pedro, J. R. N. *Lapachol: safety and efficacy in therapy.* Brazilian Journal of Pharmacognosy, 2002;

Fabri, R. L.; Nogueira, M. S.; Braga, F. G.; Coimbra, E. S.; Scio, S. *Mitracarpus frigidus aerial parts exhibited potent antimicrobial, antileishmanial, and antioxidant effects.* BioresourceTechnology, 100, 428-433, 2009;

Farah, A.; Donangelo, C. M. *Phenolc compounds in coffee.*

Brazilian Journal of Plant Physiology, v. 18, n. 1, p. 23 - 36, 2006;

Ferreira, R. A.; Oliveira, A. B. M.; Gualberto, S. A.; Vitor, R. W. *Activity of natural and synthetic naphthoquinones against Toxoplasma gondii in vitro and in murine models of infection. Parasite 9:* 261-269, 2002;

Floch, H. *Leishmania tropica guyanensis n.sp. agent de la leishmaniose tégumentaire des Guyanes et de Amérique Centrale.* Archives de l'institut Pasteurde la Guyanefrangaise et de l'inini, 328, 1-4, 1954;

Frézard, F.; Martins, P. S.; Barbosa, M. C. M.; Pimenta, A. M. C.; Ferreira, W. A.; De Melo, J. E.; Mangrum, J. B.; Demicheli, C. *New insights into the chemical structure and composition of the pentavalent antimonial drugs, meglumine antimonite and sodium stibogluconate.* Journal of inorganic Biochemistry, 102: 656-665, 2008;

Galliano, A. G. O *mètodo cientifico: teoria e pratica.* Sao Paulo: Harbra, 1986;

Garcia, E. A. C. *Manual de sistematizaçao e normalizaçao de documentos técnicos.* Sao Paulo: Atlas, 1998;

Gardener, P. J.; Chance, M. L.; Peters, W. *Biochemical taxonomy of Leishmania. II: Electrophoretic variation of malate dehydrogenase.* Annals of Tropical Medicineand Parasitology, Sep;68(3):317-325, 1974;

GIL, A. C. *Como elaborar projetos de pesquisa.* Sao Paulo: Atlas, 1991 ;

Gonzales, A. G.; Reyes, R. E.; Mato, C. & Braun, A. M. E. *Three lignans from blupeurum salicifolium.* Phytochemistry, v. 29, p. 1981 - 1983, 1990;

Gottlieb, 0. R. *Lignoids from Amazonian plants: biological and chemical investigations.* ActaAmazonica, v. 18, p. 333-344, 1988;

Grimaldi Jr., G.; Tesh, R. B.; McMahon-Pratt, D. *A review of the geographic distribution and epidemiology of leishmaniasis in the New World.* American Journal of Tropical Medicine and Hygiene, 41: 687-725, 1989;

Harbone, J.B. & Baxter, H. *Phytochemical dictionary: a handbook of bioactive compounds from plants.* Taylor & Francis, London, 1995;

Hartwell, J. L. & Scherecker, A. W. *The chemistry of pedophyllum Fort.* Chemestry Organic Naturst, v. 15, p. 83-116, 1958;

Hazra, B.; Sarkar, R.; Bhattacharyya, S.; Ghosh, P. K.; Chel, G.; Dinda, B. *Synthesis of plumbagin derivatives and their inhibitory activities against Ehrlich ascites carcinoma in vivo and Leishmania donovani promastigotes in vitro.* Phytotherapy Research, *16:* 133-137, 2002;

Hiroyuki, H.; Katsuhiko, T.; Yukiyoshi, T.; Kenji, M.; Takeshi, K. *Mode of antibacterial action of retrochalcones from Glycyrrhiza inflata.* Phytochemistry 48, 125-129, 1998;

Honigberg, B. M. A. *The contribution to systematic of the non-pigmented flagellates.* In: Ludvik J., Lom J., Vavra J. Profress in Protozoology. NewYork: Academic Press, p. 68, 1963;

Jorge, R.; Furtado, N. A. J. C.; Sousa, J. P. B.; Da Silva Filho, A. A.; Gregorio,

L. E.; Martins, C. H. G., et al. *Brazilian Propolis: Seasonal Variation of the Prenylated p-Coumaric Acids and Antimicrobial Activity.* Pharmaceutical Biology, 46:889, 2008;

Kayser, 0.; Kiderlen, A. F.; Laatsch, H.; Croft, S. L. *In vitro leishmanicidal activity of monomeric and dimeric naphthoquinones.* Acta Tropica, 77: 307-314, 2000;

Kemertelidze, E. P.; Tsitsishvili, V. G.; Alaniya, M. D.; Sagareishvili, T. G. *Structure-function analysis of the radioprotective and antioxidant activity of flavonoids.* Chemistry of Natural Compounds, 36, 54-59, 2000;

Kessler, P. J. A.; Kubitski, K.; Rohwer, J. C.; Bittrich, V. *In Annonaceae.*

Springer-Verlag, Berlin, vol. 2, p. 93-129, 1993;

Khan, M. R.; Kihara, M.; Moloso, A. D. *Antimicrobial activity of Terminalia complanata and Flacourtia zippelii. Fitoterapia, 73,* 744, 2002;

Lainson, R.; Shaw, J. J. *Evolution, classification and geographical distribution,* p. 1 - 120. In W. Peters, R. Killick-Kendrick. The Leishmaniasis in Biology and Medicine, Vol. 1. AcademicPress, NewYork, 1987;

Lainson, R.; Shaw, J. J. *Leishmaniasis of the New World: Toxonomic problems. British Medical* Bulletin, *28:44 - 48, 1972;*

Lakatos, E. M.; Marconi, M. A. *Metodologia do trabalho cıentıfico.* Sao Paulo: Atlas, 1995;

Leitao, D. P. S.; Da Silva Filho, A. A.; Polizello, A. C. M.; Bastos, J. K.; Spadaro, A. C. C. *Biological* & Pharmaceutical *Bulletin*, 27:1834, 2004;

Lemos, M.; Barros, M. P.; Sousa, J. P. B.; Da Silva Filho, A. A.; Bastos, J. K.; Andrade, S. F. J. *Baccharis dracunculifolia, the main botanical source of Brazilian green propolis, displays antiulcer activity.* Journal of Pharmacy and Pharmacology, 59:603-8, 2007;

Leon, L. L.; Gomes, D. C. F.; Alegrio, L. V.; Lima, M. E. F.; Araùjo, C. A. C. *Synthetic derivatives of curcumin and their activity against Leishmania amazonensis.* Arzneimittel-Forschung, 52: 120-124, 2002;

Leon, L. L.; Machado, G. M. C.; Barral, A.; Carvalho-Paes, L. E.; Grimaldi Jr, G. *Antigenic differences among Leishmania amazonensis isolates and their relationship with distinct clinical forms of the disease.* Memorias do Instituto OswaldoCruz, 87:229-234, 1992;

Lin, M.; Anderson, H.; Flavin, M. T.; Pai, Y. S. *In vitro anti-HIV activity of bioflavonoids isolated from rhus sucedanea and garciania multiflora.* Journal Natural Products, vol. 60 884-8, 1997;

Lindsey, K. L.; Budesinsky, M.; Kohout, L.; Van Staden, J. *Antibacterial activity of maytenoic acid isolated from the root-bark of Maytenus senegalensis.* South African Journal of Botany 72, 473-477, 2006;

Lorenzi H. *Arvores Brasileiras, Plantarum, Nova Odessa.* Sao Paulo, Brazil, p. 21-26, 1992;

Lorenzi, H.; Matos, F. J. A. *Plantas medicinais no Brasil: nativas e exóticas cultivadas.* 1st ed., Instituto Plantarum, Nova Odessa, *p.* 511,2002;

Lumsden, W. H. R. *Designation of strains of parasitic Protozoa.* Transactions of the Royal Society of Tropical Medicine and Hygiene 70, p. 352, 1977;

Lunardi, F.; Guzela, M.; Rodrigues, A.; Correa, R.; Eger-Mangrich, I.; Steindel, M.; Grisard, E.; Assreuy, J.; Calixto, J.; Santos, A. R. *Trypanocidal and leishmanicidal properties of substitution-containing chalcones.* Antimicrob. Agents Chemother, 47 (4), 1449-1451, 2003;

Machado, M. I.; Nilshioka, S. A.; Ferreira, M. S.; Silva, A. R.; Rocha, A.; Costa-Cruz, J. M. & Gonçalves, M. R. F. *Leishmaniasis tegumentar americana no Triângulo Mineiro e Alto Paranaiba. Revista do Centro de Ciências Biomédicas da Universidade Federal de Uberlândia,* 8:17-28, 1992;

Maregesi, S. M.; Hermans, N.; Dhooghe, L.; Cimanga, K.; Ferreira, D.; Pannecouque, C.; Berghe, D. A. V.; Cos, P.; Maes, L.; Vlietinck, A. J.; Apers, A.; Pieters, L. *Phytochemical and biological investigations of Elaeodendron schlechteranum.* Journal of Ethnopharmacology, 129, 319-326, 2010;

Maregesi, S. M.; Ngassapa, 0. D.; Pieters, L.; Vlietinck, A.J. *Ethnopharmacological survey of the Bunda district, Tanzania: plants used to treat infectious diseases.* Journal of Ethnopharmacology 113, 457-470, 2007;

Maregesi, S. M.; Pieters, L.; Ngassapa, 0. D.; Apers, S.; Vingerhoets, R.; Cos, P.; Vanden Berghe, D. A.; Vlietinck, A. J. *Screening of some Tanzanian medicinal plants from Bunda district for antibacterial, antifungal and antiviral activities.* Journal of Ethnopharmacology 119, 58-66, 2008;

Maregesi, S. M.; Van Miert, S.; Pannecouque, C.; Feiz Haddad, M. H.; Hermans, N.; Wright, C. W.; Vlietinck, A. J.; Apers, S.; Pieters, L. *Screening of Tanzanian medicinal plants against Plasmodium falciparum and Human Imumunodeficiency Virus.* Planta Medica 76, 195-201,2010;

Marzochi, M. C. A. *Leishmaniases no Brasil (The Tegumentary Leishmaniases). Brazilian Journal of Medicine, 63 (5/6): 81-105, 1992;*

Miert, S. V.; Dyck, S. V; Schmidf, T. J.; Brun, R.; Vlietinck, A.; Lumiere, G. & Pieters, L. *Antileishmanial activity, cytotoxicity and QSAR analysis of synthetic dihydrobenzofuran lignans and related benzofurans.* Bioorganic & Medicinal Chemistry, 13,66- 669, 2005;

Missima, F.; Da Silva Filho, A. A.; Nunes, G. A.; Bueno, P. C. P.; Sousa, J. P. B.; Bastos, J. K., et al. *Effect of Baccharis dracunculifolia D.C*

(Asteraceae) extracts and its isolated compounds on macrophage activation. Journal of Pharmacy and Pharmacology, 59:463, 2007;

Montenegro, H.; Gutiérrez, M.; Romero, L. I.; Ortega-Barria, E.; Capson, T. L.; Rios, L. C. *Aporphine Alkaloids from Guatteria spp. With Leishmanicidal Activity.* Planta Medica, 69, 677-679, 2003;

Morais, S. K. R. *Chemical constituents of some Myristicaceae species.* Institute of Chemistry - Postgraduate Program in Chemistry. University of Sao Paulo, 2008, 2009;

Morais, S. K. R.; Teixeira, A. F.; Torres, Z. E. S.; Nunomura, S. M.; Yamashiro- Kanashiro, E. H.; Lindoso, J. A. L. &Yoshida, M. *Biological Activities of Lignoids from Amazon Myristicaceae Species: Virola michelii, V. mollissima, V. pavonis and Iryanthera juruensis.* Printed in Brazil - ©2009 Brazilian Chemical Society. Journal of the Brazilian Chemical Society, Vol. 20, No. 6, 1110-1118, 2009;

Moulis, C.; Pelisier, J.; Bamba, D.; Fourasté, L. *Pentalogin, antifungal naphthoquinoid pigment from Mitracarpus scaber.* Second International Congress on Ethnopharmacology, Uppsala, Sweden. The Swedish Academy of Pharmaceutical Sciences, Stockholm, 1992;

Muelas-Serrano, S.; Nogal, J. J.; Martinez Diaz, R. A.; Escario, J. A.; Martinez Fernandez, J.R.; Gomez-Barrio, A. *In vitro screening of American plant extracts on Trypanosoma cruzi and Trichomonas vaginalis.* Journal of Ethnopharmacology, v.71,p. 101-107, 2000;

Murayama, T.; Eizuru, Y.; Yamada, R.; Sadanari, H.; Matsubara, K.; Rukung, G.; Tolo, F.M.; Mungai, G. M.; Kofi-Tsekpo, M. *Anticytomegalovirus activity of pristimerin, a triterpenoid quinone methide isolated from Maytenus heterophylla (Eckl. Zeyh.).* Antiviral Chemistry & Chemotherapy 18, 133-139, 2007;

Muzitano, M. F.; Tinoco, L. W.; Guette, C.; Kaiser, C. R.; Rossi-Bergmann, B.; Costa, S. S. *The antileishmanial activity assessment of unusual flavonoids from Kalanchoe pinnata.* Phytochemistry, 67, 2071-2077, 2006;

Natarajan, S.; Paily, P. P. *Effect of topical Hydrocotyle asiatica in psoriasis.*

Indian Journal of Dermatology, 18: 82-85, 1973;

Neal, R. A.; Miles, R. A. *Indirect haemagglutination test for Chagas' disease, with a simple method for survey work.* Revista do Instituto de Medicina Tropical de Sao Paulo, 1987;

Neves, D. P. *Parasitologia* humana/David Pereira Neves. - 10ª Ed. - Sao Paulo: Editora Atheneu, 2002;

Nkunya, M. H. H.; Waibel, R.; Achenback, H. *Three flavonoids from the stem bark of the antimalarial Uvaria dependens. Phytochemistry, 34*, 853, 1993;

Oliveira, A. J. B.; Koike, L.; Shepherd, S. K. L.; Reis, F. A. M. *Callus culture of Aspidosperma ramiflorum Muell.* Arg.: growth and alkaloid production. *Acta Scientiarum 23:* 609-612, 2002;

Pal, S.; Chaudhuri, A. K. N. *Studies on the anti-ulcer activity of a Bryophyllum pinnatum leaf extract in experimental animals.* Journal of Ethnopharmacology, 33, 97-102, 1991;

Pelletier, F. J. *Plato on Not-Being.* Midwest Studies in Philosophy, pp. 35-65, 1983;

Perez-Victoria, J. Ma.; Perez-Victoria, F. J.; Castanys, S.; Gamarro, F.

Therapeutic strategies and molecular bases of drug resistance to leishmaniasis. Biojournal. 5: 1-18, 2006;

Pessoa, S. P. *Classification of leishmaniases and species of the genus Leishmania.* Arguivo Faculdade HIG Sao Paulo, 26:41-50, 1961;

Pessoa, S. P.; Martins, A. V. *Parasitologia médica.* 11. ed. Rio de Janeiro:

Guanabara Koogan, ch. 7, p. 77-118, 1982;

Pessoa, S. P.; Pestana, B. R. *Natural infection of Phebotomus migonei by leptomonas forms, probably of Leishmania braziliensis.* Acta Mèdica, 1941;

Pontin, K.; Da Silva Filho, A. A.; Santos, F. F.; Silva, M. L. A.; Cunha, W. R.;

Nanayakkara, N. P. D., et al. *In vitro and in vivo antileishmanial activities of a Brazilian green propolis extract.* Parasitology Research, 103:487, 2008;

Pratt, D. M.; David, J. R. *Monoclonal antibodies that distinguish between New World species of Leishmania.* Nature. Jun 18; 291(5816): 581-583, 1981;

Queiroz, E. F.; Roblot, F.; Cavè, A.; Paulo, M. Q.; Fournet, A. *Pessoine and spinosine, two catecholic berbines from Annona spinescens.* Journal of Natural Products, 59: 438-440, 1996;

Rahman, M. M.; Lopa, S. S.; Sadik, G.; Islam, R.; Khondkar, P.; Alam, A. H. M.

K.; Rashid, M. A. *Antibacterial and cytotoxic compounds from the bark of Cananga odorata.* Fitoterapia, *76*, 758, 2005;

Rates S. M. K. *Plants as source of drugs.* Toxicon, vol. 39, issue 5, p. 603-613, May, 2001;

Rauen, F. J. *Elementos de iniciaçâo à pesquisa.* Rio do Sul-SC: Nova Era, 1999;

Reis, F. A. M.; Marques M. F. S.; Filho, H. L. F.; Kato, L. *Indole alkaloids from Aspidosperma ramiflorum.* Phytochemistry 41: 963-967, 1996;

Ribeiro, J. E. L. S.; Maas, P. J. M.; Maas, H.; Miralha, J. M.; Webber, A.; Ribeiro, J. E. L. S.; Hopkins, M. J. G.; Vicentini, A.; Sothers, C. A.; Costa, M. A. S.; Brito, J. M.; Souza, M. A. D.; Martins, L. H. P.; Lohmann, L. G.; Assunçâo, P. A. C. L.; Pereira, E. C.; Silva, C. F.; Mesquita, M. R.; Procopio, L. C. *In Annonaceae.* INPA: Manaus, p. 121-135, 1999;

Rocha, L. G.; Almeida, J. R. G. S.; Macêdo, R. O.; Barbosa-Filho, J. M. Rocha, L. G.; Almeida, J. R. G. S.; Macedo, R. O.; Barbosa-Filho, J. M. *A review of natural products with antileishmanial activity.* Phytomedicine 12: 514-535, 2005;

Rogers, D. *et al. Cachexial fever in India associated with Cunninham-Leishman- Donovan bodies. British Medical Journal,* n◦ 2, pp. 644-5, 1915;

Ross, R. *Further Notes on Leishmania's bodies.* British Medical Journal 11:1401, 1903;

Rossi-Bergmann, B.; Costa, S. S.; Borges, M. B. S.; Da Silva, S. A. G.; Noleto, G. R.; Souza, M. L. M.; Moraes, V. L. G. *Immunosuppressive effect of the aqueous extract of Kalanchoe pinnata in mice.* Phytotherapy Research, vol. 8, issue 7, p. 399-402, November, 1994;

Sahpaz, S.; Bories, C.; Loiseau, P. M.; Cortes, D.; Hocquemiller, R.; Laurens, A.; Cave, A. *Cytotoxic and antiparasitic activity from Annona senegalensis seeds.* Planta Medica, 60: 538-540, 1994;

Sanogo, R.; Germano, M. P.; De Pasquale, R.; Keita, A.; Bisignano, G. *Selective antimicrobial actitivties of Mitracarpus scaber Zucc against Candida albicans and Staphylococcus sp.* Phytomedicine 2, 265-268, 1996;

Santos, D. O.; Coutinho, C. E. R.; Madeira, M. F.; Bottino, C. G.; Vieira, R. T.; Nascimento, S. B.; Bernardino, A.; Bourguignon, S. C.; Corte-Real, S.; Pinho, R. T.; Rodrigues, C. R.; Castro, H. C. *Leishmaniasis treatment - a challenge that remains: a review.* Parasitology Research. 103, 1-10, 2008;

Schinor, E. C.; Salvador, M. J.; Tomaz, J. C.; Pral, E. M. F.; Alfieri, S. C.; Albuquerque, S.; Ito, I. Y.; Dias, D. A. *Biological activities and chemical composition of crude extracts from Chresta exsucca.* Brazilian Journal of Pharmaceutical Sciences, vol. 42, n. 1, jan./mar., 2006;

Schlecht, E.; Richter, H.; Fernàndez-Rivera, S.; Becker, K. Gastrointestinal passage of Sahelian roughages in cattle, sheep and goats, and implications for livestock-mediated nutrient transfers. Animal Feed Science and Technology, *137, 93-114, 2007;*

Schnur. L. F. & Zuckerman, A. *Leishmanial excreted factor (EF) serolypes in Sudan, Kenya and Ethiopia.* Amlals of Tropical Medicint and Parasitololr, 71. 273-29-l, 1977;

Schnur. L. F.; Zuckerman, A. & Greenblatt, G. L. *Leishmanial serotypes as distinguished by the gel diffusion of factors excreted in vitro and in vivo.* Journal of Medical Sciences, vol. 8, p. 932-942, 1972;

Scott, P.; Pearce, E.; Natovitz, P.; Sher, A. *Vaccination against cutaneous leishmaniasis in a murine model. II. Immunologic properties of protective and nonprotective subfractions of soluble promastigote extract.* Journal of Immunology, 139:3118-25, 1987;

SHAHIDI, F.; NACZK, M. *Food Phenolics: sources, chemistry, effects and applications.* Lancaster: Technomic, p. 331, 1995;

Shaw J. J., Lainson R. *Leishmaniasis in Brazil: Some observations on intradermal reactions to different trypanosomatid antigens of patients suffering from cutaneous and mucocutaneous leishmaniosis.* Transactions of the Royal Society of Tropical Medicine and Hygiene 69:323-335, 1976;

Shin, J. S.; Kim, K. S.; Kim, M. B.; Jeong, J. H.; Kim, B. K. *Synthesis and hypoglycemic effect of chrysin derivatives.* Bioorganic& MedicinalChemistry Letters. 9, 869-874, 1999;

Silveira, T. G. V.; Arraes, S. M. A. A.; Pereira, D.S. et al. *Evaluation of the indirect immunofluorescence reaction for American tegumentary leishmaniasis in patients from the northwestern region of the state of Paranâ-Brazil.* Revista Unimar, 1987;

Simoes, L. M. C.; Gregorio, L. E.; Da Silva Filho, A. A.; De Souza, M. L.;

Azzolini, A. E. C. S.; Bastos, J. K., et al. *Effect of Brazilian green propolis on the production of reactive oxygen species by stimulated neutrophils.* Journal of Ethnopharmacology 94: 59-65, 2004;

Simpson, L. *The mitochondrial genome of Kinetoplastid protozoa: genomic organization, transcription, replication and evolution.* Annual Review of Microbiology 41, 363-382, 1987;

Sousa, J. P. B.; Bueno, P. C. P.; Gregorio, L. E.; Da Silva Filho, A. A.; Furtado, N. A. J. C.; Sousa, M. L., et al. *A reliable quantitative method for the analysis of phenolic compounds in Brazilian propolis by reverse high performance liquid chromatography.* Journal of Separation Science, *30:* 2656-2665, 2007;

Teixeira, M. J.; De Almeida, Y. M.; Viana, J. R.; Holanda, J. G.; Rodrigues, T. P.; Prata, J. R. C.; Coelho, I. V. B.; Rao, V. S.; Pompeu, M. M. L. *In vitro and in vivo leishmanicidal activity of 2-hydroxy-3-(3-methyl-2-butenyl)-1,4- naphthoquinone (lapachol).* Phytotherapy Research, Volume 15, Issue 1, p. 4448, February, 2001;

Tempone, A. G.; Treiger, B. S. E.; Andrade Jr., H. F.; Amorim, G. N. C.; Yogi, A.; Salerno, C. C.; Bachiega, D.; Lupo, F. N.; Bonotto, S. V.; Fischer, D. C. H. *Antiprotozoal activity of Brazilian plant extracts from isoquinoline alkaloid- producing families.* Phytomedicine, 12: 382-390, 2005;

Thiem, D. A.; Sneden, A. T.; Khan, A. I.; Tekwami, B. L. *Bisnortriterpenes from Salacia madagascariensis.* Journal of Natural Products 68, 251-254, 2005;

Tona, L.; Cimanga, R. K.; Mesia, K.; Musuamba, C. T.; De Bruyne, T.; Apers, S.; Hermans, N.; Van Miert, S.; Pieters, L.; Totté, J.; Vlietinck, A. J. *In vitro antiplasmodial activity of extracts and fractions from seven medic inal plants used in the Democratic Republic of Congo.* Journal of Ethnopharmacology, 93, 27-32, 2004;

Tona, L.; Mesia, K.; Ngımbı, N. P.; Chrimwami, B.; Okond'Ahoka; Cimanga, K.;

De Bruyne, T.; Apers, S.; Hermans, N.; Totté, J.; Pieters, L.; Vlietlnck, A. J. *In vivo antimalarial activity of Cassia occidentalis. Morinda morindoides and Phyllanthus niruri.* Annals of Tropical Medicine and Parasitology, 95, 47-57, 2001;

Tona, L.; Ngimbi, N. P.; Tsakala, M.; Mesia, K.; Cimanga, K.; Apers, S.; De Bruyne, T.; Pieters, L.; Totté, J.; Vlietinck, A. *Antimalarial activity of 20 crude extracts from nine African medicinal plants used in Kinshasa, Congo.* Journal of Ethnopharmacology, 68, 193-203, 1999;

Trivinos, A. N. S. *Introdugao à pesquisa em ciências sociais: a pesquisa qualitativa em educaçâo.* Sao Paulo: Atlas , 1987;

Valderrama, J. C. M. *Distribution of flavonoids in the Myristicaceae.* Phytochemistry, v. 55, n. 6, p. 505 -511,2000;

Veiga-Junior V. F. *Estudo do consumo de plantas medicinais na Regiâo Centro-Norte do Estado do Rio de Janeiro: aceitação pelos profissionais de saúde e modo de uso pela populaçâo.* Revista Brasileira de Farmacognosia, *18:* 308-313, 2008;

Veiga-Junior V. F.; Mello J. C. P. *Monographs on medicinal plants.* Revista Brasileira de Farmacognosia, *18:* 464-471, 2008;

Vianna, G. O. *Anais do 7° Congresso Brasileiro de Medicina e Cirurgia. 4,* 426, 1912;

Viannia, G. *On a species of Leishmania.* Brasil Mèdico, 25:411, 1991;

Vickerman, K. *The diversity of the Kinetoplastid flagellates. In: Biology of the Kinetoplastida (W. H. R. Lumdsen and D. A. Evans).* Vol. 1, p. 1-34. New York and London: Academic Press, 1976;

W. H. O. *Control of leishmaniases: Report of WHO expert committee.* Technical report series, 793, Geneva; 1990;

W. H. O. *World Health Organization. Control of Chagas Disease.* Technical report series, 905, p. 1-109, 2002;

Watt, J. M.; Breyer-Brandwijk, M. G. *Medicinal and Poisonous Plants of Southern and Eastern Africa. Livingstone.* E & S Livingstone Ltd, Edinburgh, London, p. 1457, 1962;

Wilrth, D. F.; Rogers, W. O.; Barker, R.; Dourado, H.; Suesebang, L.; Albuquerque, B. *Leishmaniasis and malaria: DNA probes for diagnosis and epidemiologic analysis.* Annals New York Academic Society, New York, v. 569, p. 183-192, 1990;

Wilson, M. E. & Pearson, R. D. *Stage-specific variations in lectin binding to Leishmania donovani.* Infection and Immunity, 46 (1): 128-134, 1990;

Woodcock, H. M. *The life-cycle of Cystobia irregularis (Minch), together with observations on other "neogamous" gregarines.* Quarterly Journal of Microscopical Science, 50:1 - 100, 1906;

Yadav, N. P.; Dixit, V. K. *Hepatoprotective activity of leaves of Kalanchoe pinnata Pers.* Journal of Ethnopharmacology, 86, 197-202, 2003;

Yamagishi, T.; Haruna, M.; Yan, X. Z.; Chang, J. J.; Lee, K. H. *Antitumor agents, 1101,2 Bryophyllin B, a novel potent cytotoxic bufadienolide from Bryophyllum pinnatum.* Journal of Natural Products, 52, 1071- 1079, 1989;

Zuanazzi, J. A. S.; Montanha, J. A.; Simoes, C. M. O.; Schenkel, E. P.; Gosmann, G.; Mello, J. C. P.; Mentz, L. A.; Petrovick, P. R. Flavonoids. *In: Farmacognosia da planta ao medicamento.* Porto AlegreZFlorianopolis: Editora da UFRGSZEditora da UFSC, chapter 23, p. 577 - 614, 2004.

MIX
Papier aus verantwortungsvollen Quellen
Paper from responsible sources
FSC® C105338

Printed by Books on Demand GmbH, Norderstedt / Germany